DOT&E STRATEGY IMPLEMENTATION PLAN – 2023

NICKOLAS GUERTIN

NIMBLE BOOKS LLC: THE AI LAB FOR BOOK-LOVERS

~ FRED ZIMMERMAN, EDITOR ~

Humans and AI making books richer, more diverse, and more surprising.

PUBLISHING INFORMATION

ISBN: 978-1-60888-222-9

AI-GENERATED KEYWORD PHRASES

DOT&E Strategy Implementation Plan;
strategic intent;
key actions;
DOT&E Strategy Update;
test the way we fight;
accelerate the delivery of weapons that work;
foster an agile and enduring T&E workforce;
improve the survivability of DOD in a contested environment;
pioneer T&E of weapon systems built to change over time;
accurate evaluation of warfighting capabilities;
digital technologies and data management solutions;

PUBLISHER'S NOTES

Nimble Books makes this document available in print because the Department of Defense's Director of Operational Test and Evaluation (DOT&E) is one of the few things (other than hypersonics and nuclear-powered satellites) that can kill big weapons programs! Careful study of this document is well worth the investment of time for the student of defense policy.

This annotated edition illustrates the capabilities of the AI Lab for Book-Lovers to add context and ease-of-use to manuscripts. It includes several types of abstracts, building from simplest to more complex: TLDR (one word), ELI5, TLDR (vanilla), Scientific Style, and Action Items; essays to increase viewpoint diversity, such as Grounds for Dissent, Red Team Critique, and MAGA Perspective; and Notable Passages and Nutshell Summaries for each page.

ANNOTATIONS

ABSTRACTS

TL;DR (ONE WORD)

Implementation.

EXPLAIN IT TO ME LIKE I'M FIVE YEARS OLD

This document is like a plan that tells us how to make sure the things the military uses in wars are really good and work well. It has five important things it wants to focus on: testing how we fight, making weapons that work faster, having people who test things that are good at their job, making sure the military can survive in a tough situation, and testing weapons that can change over time.

The plan says we need to check how good the fighting things are and use technology and data to do it. Smart!

TL;DR (VANILLA)

The DOT&E Strategy Implementation Plan outlines the key actions for implementing the DOT&E Strategy Update. It focuses on five strategic pillars and emphasizes accurate evaluation of warfighting capabilities, digital technologies, and data management solutions. It also highlights the importance of integrating T&E in model-based engineering and conducting mission-based risk assessments. The plan aims to leverage digital technologies to improve T&E processes.

SCIENTIFIC STYLE

The DOT&E Strategy Implementation Plan outlines the strategic intent and key actions for implementing the DOT&E Strategy Update. The plan focuses on five strategic pillars, including testing the way we fight, accelerating weapon delivery, fostering an agile T&E workforce, improving DOD survivability in a contested environment, and pioneering T&E of evolving weapon systems. The plan emphasizes accurate evaluation of warfighting capabilities, digital technologies and data management solutions, cyber and electromagnetic spectrum survivability evaluation, integration of T&E in model-based engineering, and mission-based risk assessments. The document provides specific actions and deliverables for each pillar and proposes a phased approach for implementation. The overall goal is to leverage digital technologies and data management solutions to enhance the effectiveness and efficiency of T&E processes.

ACTION ITEMS

Develop and implement a comprehensive training program for T&E workforce to enhance their skills and knowledge in digital technologies, AI/machine learning, and data management solutions.

Establish partnerships with industry and academia to leverage their expertise and resources in developing and implementing advanced T&E methodologies.

Conduct regular assessments of cyber and electromagnetic spectrum survivability of DOD systems to identify vulnerabilities and develop mitigation strategies.

Integrate T&E activities into the model-based engineering process to ensure that testing is conducted throughout the system development lifecycle.

Develop a standardized framework for conducting mission-based risk assessments to prioritize testing efforts based on operational needs.

VIEWPOINTS

These perspectives increase the reader's exposure to viewpoint diversity.

GROUNDS FOR DISSENT

Lack of emphasis on human expertise: A member of the organization might dissent from this report because they believe that it places too much emphasis on digital technologies, AI/machine learning, and data/knowledge management solutions. They may argue that these technological advancements cannot fully replace the expertise and experience of human testers in evaluating warfighting capabilities. They might believe that relying too heavily on automation and data analysis could lead to oversights or inaccuracies in the evaluation process.

Insufficient focus on real-world scenarios: Another reason for dissent could be the perception that the plan does not adequately prioritize testing in realistic operational environments. The member might argue that the document's focus on model-based engineering and mission-based risk assessments does not sufficiently capture the complexities and challenges of warfare. They may advocate for more emphasis on field testing and simulations that closely replicate real-world scenarios, as well as incorporating feedback from military personnel who have experienced combat situations.

Neglect of long-term sustainability: Some members might take issue with the plan's heavy emphasis on delivering weapons quickly without adequate consideration for their long-term sustainability. They might argue that prioritizing expedited delivery could compromise thorough testing and evaluation, leading to potential issues or failures down the line. These individuals may advocate for a more balanced approach that ensures both timely delivery and rigorous testing, with a focus on developing weapons systems that can adapt and evolve over time.

Limited consideration of ethical implications: Dissenters might also raise concerns about the plan's apparent lack of consideration for ethical implications related to T&E processes. They may argue that integrating AI/machine learning without clear guidelines or oversight could lead to unintended consequences or unethical practices. These individuals might call for a more comprehensive examination of the ethical dimensions surrounding T&E, including issues such as privacy, bias, accountability, and potential misuse of technology.

Inadequate allocation of resources: Members might dissent if they feel that the plan does not allocate sufficient resources, both financial and human, to effectively implement the outlined actions and deliverables. They may argue that without adequate support, the plan's ambitious goals and objectives could remain unattainable or result in subpar outcomes. These individuals might advocate for a more realistic assessment of available resources and a prioritization of critical needs to ensure successful implementation.

It is important to note that these dissenting views are hypothetical and do not reflect actual criticisms of the DOT&E Strategy Implementation Plan.

RED TEAM CRITIQUE

The DOT&E Strategy Implementation Plan is a comprehensive document that outlines the strategic intent and key actions for implementing the DOT&E Strategy Update. The plan focuses on five strategic pillars, which provide a strong foundation for driving improvements in T&E processes. However, there are several areas where further analysis and clarification are needed.

Firstly, while the plan emphasizes the need for accurate evaluation of warfighting capabilities, it doesn't provide specific details on how this will be achieved. It would be beneficial to include more information on the methodologies and tools that will be used to assess these capabilities. Additionally, clear metrics should be established to measure progress in this area.

Furthermore, the implementation of digital technologies and data management solutions is mentioned as a key action in the plan. However, there is limited discussion on how these technologies will be integrated into existing T&E processes. It would be valuable to include examples or case studies that demonstrate how digital technologies have been successfully utilized in other organizations or projects.

Additionally, while cyber and electromagnetic spectrum survivability are highlighted as important areas of evaluation, there is minimal detail provided on how these assessments will be conducted. It would strengthen the plan if specific methodologies or approaches were outlined to ensure comprehensive evaluations of cyber and electromagnetic spectrum vulnerabilities.

Moreover, although model-based engineering integration and mission-based risk assessments are mentioned as important components of T&E processes, they are not given sufficient attention in the plan. These areas deserve more detailed explanations of their significance within T&E operations and strategies for their successful implementation.

Lastly, while AI/machine learning and data/knowledge management solutions are acknowledged as means to improve effectiveness and efficiency within T&E processes by leveraging digital technologies; no concrete plans regarding their utilization are presented within this document. Providing a roadmap for incorporating these technologies into current practices could enhance understanding among stakeholders about expectations during deployment phases.

In conclusion, the DOT&E Strategy Implementation Plan serves as a solid starting point for implementing improvements in T&E processes. However, it would benefit from further clarification and expansion in several areas. By addressing the issues raised in this critique, the plan could provide more actionable guidance for achieving its strategic goals.

MAGA PERSPECTIVE

This document is just another example of the deep state's agenda to undermine President Trump and his America First policies. The DOT&E Strategy Implementation Plan is filled with buzzwords and empty promises that have nothing to do with putting American interests first. Instead of focusing on important issues like border security or the economy, this plan wastes our time and resources on testing methods that are completely irrelevant to real-world combat situations.

The five strategic pillars outlined in this plan are nothing more than a smokescreen for the globalist elites who want to weaken our military and make us beholden to foreign nations. Accelerating the delivery of weapons may sound good on paper, but what good are these weapons if we don't have a secure border or a strong manufacturing base? Our priority should be protecting American jobs and industries, not rushing to produce weapons that may or may not work.

Furthermore, the emphasis on digital technologies and data management solutions is just another way for the deep state to control and surveil us. We don't need AI/machine learning or invasive cyber programs to improve our T&E processes; we need common sense and a focus on protecting American interests. This plan fails to address the real threats facing our nation, such as China's economic dominance or illegal immigration.

In addition, integrating T&E in model-based engineering is just another attempt by the globalists to push their climate change agenda. They want to use our military as a tool for their environmental goals instead of focusing on defending our country. This document completely ignores the importance of national sovereignty and puts international cooperation above American exceptionalism.

Overall, this DOT&E Strategy Implementation Plan is yet another example of how the deep state is working against President Trump and his supporters. It prioritizes globalism over American interests and fails to address the real challenges facing our nation. As true patriots, we must reject this plan and demand that our leaders prioritize America First policies that will truly make our country great again.

PAGE-BY-PAGE SUMMARIES

collaboration with government, industry, academia, and allies. The future T&E enterprise will be empowered by digital tools and continuous learning to support the warfighter and stay ahead of adversaries.

NOTABLE PASSAGES

BODY-2   *"To support delivery of the world's most advanced warfighting capabilities at the speed of need, the T&E community must work together and aggressively expand its core capabilities. This Implementation Plan (I-Plan) provides the framework to achieve the vision articulated in the DOT&E Strategy Update 2022."*

BODY-3   *"To highlight the criticality of the need to think and act like an Enterprise."*

BODY-4   *"I am confident the I-Plan's five strategic pillars will empower T&E investment and improvements in the emerging opportunity and defense-specific areas of hypersonic weapons, electronic warfare, nuclear modernization, trusted AI, and multi-domain operations."*

BODY-5   *"The rapidly increasing military capabilities of China and Russia, alongside the accelerated pace of technological change, calls for transformation. Artificial intelligence, ubiquitous sensors, unmanned systems, and long-range precision weapons make contested environments more lethal than ever and will transform how the Navy will fight in the future."*

BODY-6   *"The Air Force Science & Technology 2030 Strategy emphasizes the need for improvements in such areas as hypersonics, artificial intelligence, stealth technologies, and other disruptive inventions to ensure our nation's defense. As our adversaries work towards reaching technological parity, we must set the pace to drive towards the vision of dominating time, space, and complexity across all operating domains."*

BODY-11   *"To enable efficient and structured modernization and sustainment of existing range capabilities while also transforming the ranges to meet the demands of the future, it is important to have an accurate and common picture (a dashboard) of existing and required, future range capabilities. It will be equally important to ensure this common picture is digitized, and transparent to key T&E stakeholders to enable collaboration in developing joint/interoperable solutions, avoiding redundancies while increasing capability delivery and efficiencies."*

BODY-12   *"DOT&E intends to establish a multi-disciplinary team comprised of government experts, consultants, academia, industry partners, small businesses, intelligence community and international partners to demonstrate progress in meeting the desired end state for Pillar 1."*

BODY-13   *"With the emergence of joint all-domain command-and-control solutions and the concept of kill webs, it is important to define the process and the required T&E tools that would effectively measure the success rates of mission threads, concepts, and solutions."*

BODY-14   *"Success will depend on collaboration with USD(R&E) and its Mission Engineering Team, the Combatant Commands, large force exercise leads, USD(A&S), the Intelligence Community, and existing Joint All Domain Command & Control cross-functional teams."*

BODY-15   *"As data-driven and complex systems continue to proliferate, it is important to develop T&E data and interface standards, stores, and platforms to ensure that the data are credible, trustworthy, available, and secure across the T&E enterprise."*

BODY-16   *"DOT&E intends to establish a multi-disciplinary team comprised of government experts, consultants, academia, industry partners, small businesses, intelligence community and international partners to demonstrate progress in meeting the desired end state for Pillar 2."*

BODY-17   "Modern model-based engineering and adaptive inference processes offer integrated, holistic approaches to generating and managing knowledge of system performance throughout the lifecycle."

BODY-18   "DOT&E intends to establish a multi-disciplinary team comprised of government experts, consultants, academia, industry partners, small businesses, intelligence community and international partners to demonstrate progress in meeting the desired end state for Pillar 2."

BODY-19   "Seamless integration of multiple systems and technologies working together across multiple domains introduces a potential for vulnerabilities that cannot be evaluated one system or one threat at a time. As discussed in the first (Test the Way We Fight) pillar, testing must consider the mission thread, specifically the composition of weapon systems, networks, critical infrastructure, equipment, and Tactics, Techniques, and Procedures, in accordance with the Defense Planning Guidance, to fully account for the attack surface. A rapid and accurate mission-based risk assessment would define specific steps to enhance mission assurance and identify the defenses required against threats to those missions."

BODY-20   "DOT&E intends to establish a multi-disciplinary team comprised of government experts, consultants, academia, industry partners, small businesses, intelligence community and international partners to demonstrate progress in meeting the desired end state for Pillar 3."

BODY-21   "Our challenge is to evaluate cyber-physical systems against advanced cyber and electromagnetic spectrum threats at scale and speed. Attack surfaces are growing exponentially, reaching into supply chains, software factories, and pipelines; the electromagnetic spectrum; and an array of cloud solutions. We therefore must aggressively pursue verified and validated digital tools and transformative technologies to manage cyber, electromagnetic spectrum and advanced kinetic threat survivability T&E and assess the effectiveness of countermeasures and other self-defense solutions."

BODY-22   "DOT&E intends to establish a multi-disciplinary team comprised of government experts, consultants, academia, industry partners, small businesses, intelligence community and international partners to demonstrate progress in meeting the desired end state for Pillar 3."

BODY-23   "Space is increasingly congested and highly contested, with a broad array of rapidly evolving threats. Reliance on space-based capabilities has sharpened DOD's – and our adversaries' – focus on deploying both offensive and defensive weapons in space. Because the DOD must operate in this contested environment, the T&E enterprise must be ready to accurately evaluate space-based and space-dependent systems' operational performance, to include survivability against current and anticipated threats."

BODY-24   "The combination of new domains and operational constraints makes verified, validated and accredited digital technologies the necessary, practical approach for development and T&E of certain systems where live fire T&E is not possible or practical. For example, digital twins that can be subjected to repeated cyber-attacks – as the system itself, the threats it will face, and adversary tactics, techniques, and procedures change over time – will help developers and program managers improve system cyber survivability at the required pace."

BODY-25   "Develop a process and criteria that need to be met to effectively leverage the use of digital twins in T&E to include an agile and iterative verification, validation, and accreditation process to continuously improve and update digital twins across the program lifecycle as new live data become available."

BODY-26   "Ethical and safe use of AI is necessary to reduce risks to U.S. strategic initiatives, reputation, operations, legal standing, and privacy issues."

BODY-27   "Identify areas that require further research to support adequate operational and ethical performance of AI-enabled systems (e.g., research/develop procedures for translating AI-enabled system testing into verifiable requirements that incorporate DOD AI ethical principles). Ontology of research needs summarizing the milestones and deliverables needed to evaluate the operational and ethical performance of AI-enabled systems, to include the requirements and processes for AI red teams. Defined use case of evaluation of AI-based system, identified gaps, and scoped requirements. Defined action plan to address gaps in evaluation and allocate resources accordingly. Implementation of the research agenda with pilot/demonstration case intended to close the identified T&E gaps."

BODY-28   "It is important to identify new approaches to address change propagation within software-reliant systems. The T&E community needs to influence and measure the development and cyber defense of software pipelines and factories upfront with accredited tools, techniques, and procedures. We need to embrace automated testing at every level and ensure that our rigorous standard of testing continues to occur at the speed of relevance."

BODY-30   "A structured approach for the development and sustainment of the T&E enterprise workforce will enhance workforce agility and response to emerging T&E requirements. The nature of T&E evolved drastically over the past decade, and the reemergence of Great Power conflict necessitates a thorough review and refinement of the T&E Competency Model to ensure T&E professional learning and development aligns to the enterprise's needs. Coupled with modernized competencies, the enterprise should better track and manage the T&E workforce's overall readiness in real-time to deliver improved talent management initiatives which strengthen the overall enterprise and provide an optimal T&E professional talent experience."

BODY-32   "The T&E FIT is well positioned to lead and oversee the enterprise's push to better connect learning opportunities to the T&E workforce and track learning records and outcomes against the T&E Competency Model. The T&E learning apparatus should evolve as quickly as the T&E operating environment, resulting in easily adaptable courses, content, and training and workforce demands. Continuous learning can be better incentivized and accessible by leveraging the enterprise's currently federated learning data as a strategic asset, and overall gaps in learning may be more easily identified and addressed."

BODY-34   "The success of these actions depends on the participation of T&E community, their willingness to freely share information, and contribute as able to expedite the delivery of the required T&E capabilities."

BODY-35   "To continue to deliver credible warfighting capability at the speed of need, we in the T&E enterprise must rethink the way we do business. We must become more agile, efficient and effective to adequately account for the technology disruptors as we face an inflection point in the scope, scalability, and capabilities of our infrastructure, tools, processes and workforce."

Director, Operational Test and Evaluation

DOT&E Strategy Implementation Plan – 2023

April 24, 2023

HON Nickolas Guertin

Director, Operational Test and Evaluation

Foreword

The 2022 National Defense Strategy calls for the U.S. military to achieve the multi-domain superiority necessary to deter and defeat all potential adversaries. The strategy focuses on designing, developing, and managing our forces in a manner that links our operational concepts and capabilities to achieve strategic objectives. This requires a Joint Force that is *lethal, suitable, resilient, survivable, agile, and responsive.*[1]

Test and evaluation (T&E) of Department of Defense (DOD) acquisition programs is essential to fielding and enabling this Joint Force. To drive effective and efficient T&E and avoid discovery of performance shortfalls on the battlefield, the DOT&E Strategy Update 2022 sets a course for transformative changes to T&E infrastructure, tools, processes, and workforce in concert with the ever-evolving landscape of advancing technologies and adversaries. Specifically, the strategy details the path forward to continue to enable accurate evaluation of the operational performance and limitations of the DOD to *Prevail in Conflict and Defend the Homeland*. It emphasizes the need to implement the DOD Data Management Strategy and to accelerate delivery of T&E data to acquisition decision makers to *Build Enduring Advantage*. The strategy also focuses on increasing DOD's ability to illuminate and help mitigate vulnerabilities at scale, frequency, and depth to improve survivability in a contested environment and build *Defense/Resilience*. It aligns with the DOD Responsible Artificial Intelligence Strategy and Implementation Pathway, emphasizing the need for continuous evaluation of the operational and ethical performance of artificial intelligence (AI) and similar systems designed to change over time. Lastly, and equally important, DOT&E Strategy Update 2022 focuses on fostering an agile and enduring T&E enterprise workforce to enable DOD's *Readiness and Training*.

To support delivery of the world's most advanced warfighting capabilities at the speed of need, the T&E community must work together and aggressively expand its core capabilities. This Implementation Plan (I-Plan) provides the framework to achieve the vision articulated in the DOT&E Strategy Update 2022. I invite the Service T&E and Acquisition Executives, Service Operational Test Commanders, my OSD colleagues, the Intelligence Community, the Joint Staff, and Combatant Commanders to work together to achieve the desired intent identified in this I-Plan. Our obligation to the warfighter is to convey to them, with confidence, how their equipment will perform as they plan and execute their missions. We must continue to uphold that commitment despite the challenges that lie ahead.

HON Nickolas Guertin
Director, Operational Test and Evaluation

---

[1] Italicized language in this document identifies terms used in the National Defense Strategy 2022

2 DOT&E – Strategy Implementation Plan

Think and Act Like an Enterprise

Successful implementation of the DOT&E Strategy Update will require more detailed coordination with the T&E enterprise. This Implementation Plan (I-Plan) describes the fiscal year 2023 (FY23) activities that we in DOT&E have already funded to initiate the implementation of the strategy. This I-Plan also documents the intent to achieve the following in FY23:

- **Develop a charter intended to define the resources, roles and responsibilities of key T&E strategy I-Plan stakeholders** and bring the T&E community together to identify and solve common challenges.

- **In accordance with the charter, facilitate cross-organizational discussions, increase awareness and identify/leverage ongoing efforts** that can be scaled to accelerate the capability of the T&E infrastructure, tools, processes, workforce and T&E strategies.

- **Converge on the FY24 priorities and measures** for the entire T&E enterprise and document them in the I-Plan Update in 2024.

- **Continue to provide transparency through ongoing T&E initiatives** to find efficiencies within the Services and other DOD components and look for opportunities to accelerate innovation.

Under Secretaries' Endorsement

Solving the Department's tough operational, engineering, and mission-focused challenges to stay ahead of our competitors demands collective cooperation and broad community engagement. This I-Plan provides key goals for our community to work together as a unified enterprise to bring greater efficiency, credibility, and insights throughout the T&E lifecycle of emerging technology needed to support warfighting capabilities now and in the future.

I am confident the I-Plan's five strategic pillars will empower T&E investment and improvements in the emerging opportunity and defense-specific areas of hypersonic weapons, electronic warfare, nuclear modernization, trusted AI, and multi-domain operations.

In our ever-changing and fastmoving global environment, technological advantage is not stagnant. I look forward to resolving the anticipated challenges arising from essential follow-on work to improve I-Plan execution by partnering with DOT&E, the Services, and my T&E colleagues throughout the Department.

Ms. Heidi Shyu
Under Secretary of Defense for Research and Engineering/Chief Technology Officer

This I-Plan is a crucial step in DOD moving towards more effective T&E. It outlines our path forward to sustain and improve our T&E infrastructure and workforce, made even more important given the dynamic technologies that our department is pursuing while also addressing the pacing threat environment. Moreover, it calls for closer coordination and collaboration as one T&E enterprise. The Office of the Under Secretary of Defense for Acquisition and Sustainment (OUSD(A&S)) will be a key teaming partner in the coordination and execution of the I-Plan, providing the skills and expertise required to accelerate our T&E transformation. I look forward to working with my colleagues to better integrate our T&E processes and capabilities with the acquisition processes so that we can stay ahead of the adversary.

HON Dr. William LaPlante
Under Secretary of Defense for Acquisition and Sustainment

Service Secretaries' Endorsements

The Army recognizes that the DOD operates in a dynamic environment marked by constant challenges posed by our adversaries and the most technical innovation seen since the end of the Cold War. We must ensure our adversaries cannot outrange or outpace us in traditional battlefields, or on the new frontiers of space and cyberspace. Integrated and advanced T&E is critical to helping us achieve this objective and meet the evolving threat. This I-Plan lays the groundwork to move the T&E enterprise forward and enable Army of 2030, which must acquire more advanced sensors, be secure from cyber and electronic attacks, and sustain the fight in contested environments. The Army looks forward to working with DOT&E and the other Services to implement this plan.

Ms. Christine Wormuth
Secretary of the Army

On behalf of the Navy, I am pleased to increase the collaboration and coordination across the T&E enterprise to accelerate innovation and prepare our infrastructure and workforce for the challenges of the future. The rapidly increasing military capabilities of China and Russia, alongside the accelerated pace of technological change, calls for transformation. Artificial intelligence, ubiquitous sensors, unmanned systems, and long-range precision weapons make contested environments more lethal than ever and will transform how the Navy will fight in the future.

This I-Plan provides a guiding framework to transform the way we do business and captures many of the initiatives that we in the Navy have already been pursuing. I am particularly interested in Pillar 1 and how we can reimagine how we test the way we fight. I look forward to working with my colleagues to achieve tactical and strategic wins based on this path forward.

HON Carlos Del Toro
Secretary of the Navy

I resonate with the aspirations outlined in this I-Plan – particularly the need to modernize our T&E infrastructure as highlighted in Pillar 1. The Air Force Science & Technology 2030 Strategy emphasizes the need for improvements in such areas as hypersonics, artificial intelligence, stealth technologies, and other disruptive inventions to ensure our nation's defense. As our adversaries work towards reaching technological parity, we must set the pace to drive towards the vision of dominating time, space, and complexity across all operating domains.

Improved T&E tools and processes within these areas is key to our success. It is more critical than ever that we collaborate to solve common challenges, and the Air Force looks forward to contributing to the work ahead.

HON Frank Kendall
Secretary of the Air Force

Table of Contents

Strategic Intent

The DOT&E Strategy Update 2022 outlines 5 strategic pillars and 12 lines of efforts (labeled 1.1 – 5.2 in this document), framed to achieve the following desired end state. The timeline of the desired end state will be clarified in the I-Plan Update – 2024 after a more detailed coordination with the T&E enterprise.

Desired End State

Test the way we fight

- Accurate representation of the joint, multi-domain operating environment in test
- Established process, resources and capabilities to evaluate joint warfighting capabilities and mission threads (kill webs, system-of-systems performance)

Accelerate the delivery of weapons that work

- Discoverable, accessible, and secure data repositories
- Near real-time test data analysis and assessment
- Established tools and processes that optimize integrated T&E
- Digital documentation and tracking of T&E strategies and plans

Improve the survivability of DOD in a contested environment

- Minimized mission-critical vulnerabilities and maximized defense in a contested environment
- Timely tracking and mitigation of mission-critical vulnerabilities as systems and the threats evolve

Pioneer T&E of weapon systems built to change over time

- Standardized and increased use of credible digital tools/twins in T&E
- Adequate assessment of operational and ethical performance of AI-enabled systems
- Continuous tracking and mitigation of any degradation of operational performance of DOD systems in theater

Foster an agile and enduring T&E Enterprise Workforce

- Highly skilled T&E workforce prepared to meet the toughest challenges
- Effective continuous learning program and a robust recruitment/retention plan
- Agile and innovative workforce operating model

Areas of Performance

98%

We need to identify reasonable measures of performance to adequately communicate progress to our leadership and Congress, and the value added of our initiatives. This initial I-Plan offers some suggestions and a placeholder to be updated in the I-Plan Update – 2024 after a more detailed discussion with the T&E enterprise. Coordinated discussions in FY23 should yield in defined and agree upon realistic and time-bound measures and an approach to baseline and calculate them.

Performance Areas

Test the way we fight	• Combatant Command-endorsed mission threads evaluated in T&E • Range of the future requirements identified and met
Accelerate the delivery of weapons that work	• Collected T&E data stored, secured, discoverable, and accessible to users • T&E data collection and analysis timelines • Acquisition programs effectively implementing digital technologies
Improve survivability of DOD in a contested environment	• Mission-critical vulnerabilities identified and mitigated prior to acquisition decision points and fielding • Response time to evaluate and respond to the performance against newly identified threats after fielding
Pioneer T&E of weapon systems built to change over time	• Systems for which operational performance can be digitally monitored • Acquisition programs using credible digital twins in T&E • Resolution of shortfalls necessary to test AI
Foster an agile and enduring T&E enterprise workforce	• T&E courses representing T&E core competencies • Completion of T&E training and education course offerings • Internship and scholarship placements in T&E fields

I-Plan Phased Approach

Given the scope of the Implementation Plan (I-Plan) and its reach across the DOD enterprise, we are taking a phased approach in FY23 to begin rolling out and executing already funded actions and deliverables while working on defining a T&E enterprise-level I-Plan to be delivered in FY24.

In **Phase I**, DOT&E will engage with the T&E enterprise to identify and catalog in-flight initiatives. DOT&E will simultaneously move out against already funded FY23 deliverables outlined in this document to demonstrate progress against the Strategic Intent.

In **Phase II**, in collaboration with the T&E enterprise, for each Pillar, we will provide a status assessment that includes a coordinated resolution of any redundant efforts and remaining gaps. Concurrently, we will establish the strategic foundation and governance for I-Plan success to include the process for the allocation of available resources.

Finally, in **Phase III**, gaps will be analyzed and prioritized and resourcing for T&E enterprise-wide initiatives will be aligned upon in advance of FY24.

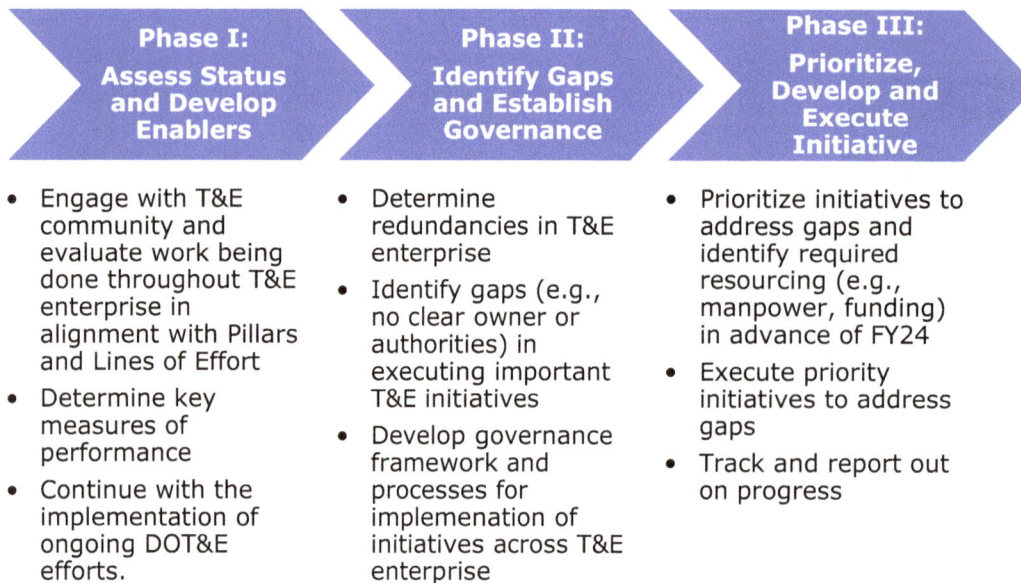

Phase I: Assess Status and Develop Enablers	Phase II: Identify Gaps and Establish Governance	Phase III: Prioritize, Develop and Execute Initiative
• Engage with T&E community and evaluate work being done throughout T&E enterprise in alignment with Pillars and Lines of Effort • Determine key measures of performance • Continue with the implementation of ongoing DOT&E efforts.	• Determine redundancies in T&E enterprise • Identify gaps (e.g., no clear owner or authorities) in executing important T&E initiatives • Develop governance framework and processes for implemenation of initiatives across T&E enterprise	• Prioritize initiatives to address gaps and identify required resourcing (e.g., manpower, funding) in advance of FY24 • Execute priority initiatives to address gaps • Track and report out on progress

Key Actions

Test the way we fight

1.1 Standardize the development of a scalable and adaptive representation of the joint, multi-domain operating environment

Why This Matters

Accurate evaluation of warfighting capabilities requires an adequate representation of the theater-representative operating environment during test and training. It also requires equipment that can adequately measure technical and operational performance of emerging or fielded warfighting capabilities in that environment. The DOD has an array of test and training ranges and capabilities managed, funded, and operated by different stakeholders. To enable efficient and structured modernization and sustainment of existing range capabilities while also transforming the ranges to meet the demands of the future, it is important to have an accurate and common picture (a dashboard) of existing and required, future range capabilities. It will be equally important to ensure this common picture is digitized, and transparent to key T&E stakeholders to enable collaboration in developing joint/interoperable solutions, avoiding redundancies while increasing capability delivery and efficiencies.

Drivers: Existing and emerging transformative technologies, such as software-reliant systems; AI; hypersonic weapons; directed energy weapons; advanced kinetic threats; contested space, electromagnetic spectrum, and cyber domains; and joint, all-domain command-and-control solutions.

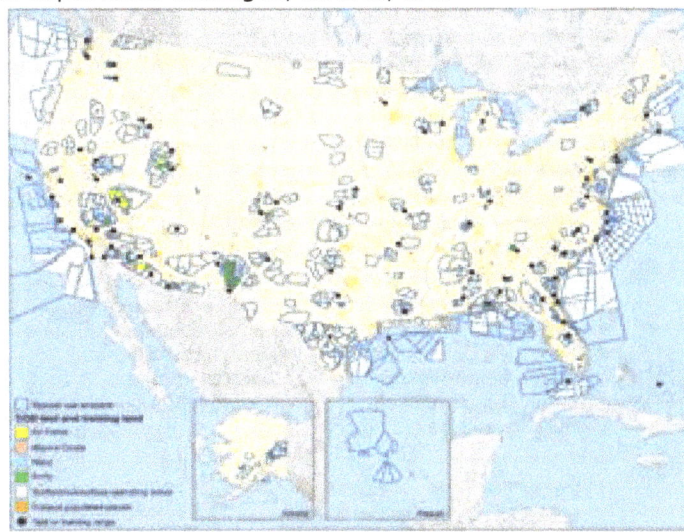

Enablers: Digital technologies, artificial intelligence, machine learning, data/ knowledge management solutions.

✓ What Success Looks Like

DOT&E intends to establish a multi-disciplinary team comprised of government experts, consultants, academia, industry partners, small businesses, intelligence community and international partners to demonstrate progress in meeting the desired end state for Pillar 1. Specifically, in FY23, DOT&E will:

- Facilitate innovation sessions, accelerate the collection of data needed to complete the actions, and generate or leverage commercial off-the-shelf solutions to deliver a concept for the range capabilities dashboard.
- Demonstrate progress in defining the concept and requirements for credible virtual/constructive all-domain modeling and simulation (M&S) solutions.
- Outline the existing range funding workflows and develop recommendations to optimize the process.

Success will depend on the cooperation of Service Test Agencies, Program Offices, and range commanders to share relevant data and inform the dashboard design solutions. Success will also depend on input and collaboration with the Test Resource Management Center (TRMC) within USD(R&E), as the primary responsible office for this line of effort.

The table below highlights Key Actions and Deliverables that DOT&E has initiated to enable this Line of Effort. T&E enterprise initiatives will be subsequently identified.

	Key Action	Deliverable	FY23 Target Deliverable	FY24 Target Deliverable	FY25 Target Deliverable
1.1.1	Develop, digitally document, prioritize, and track OT&E and LFT&E range capability requirements[2] for adequate T&E of current and emerging DOD warfighting capabilities.	Process for collecting and prioritizing T&E range capability requirements and associated cost – both physical and virtual.	Identified initial OT&E and LFT&E requirements for range of the future	Assessed gaps and defined action plan to meet requirements	Standardized process to continually update existing requirements and define and prioritize new requirements
		T&E range requirements data dashboard accessible by all T&E stakeholders.	Developed prototype[3] of range capabilities dashboard	Completed range capabilities dashboard	Refined range capabilities dashboard
1.1.2	Develop requirements and concept[4] for a data-backed, all-domain M&S environment to integrate with live, multi-domain operational testing.	Concept and requirements for an all-domain, credible M&S architecture that integrates into Live, Virtual, and Constructive Testing.	Defined initial concept and requirements	Finalized concept and requirements definition	Defined action plan to implement concept and requirements

---

[2] Requirements are defined as *inputs that govern what, how well, and under what conditions a product will achieve a given purpose*

[3] Prototype is defined as *a model suitable for evaluation of design, performance, and production potential*

[4] Concept is defined as *an abstract idea of a capability, technology, or process generalized from instances*

1.2

Implement measures, tools, and processes to efficiently evaluate mission threads, kill webs and system-of-systems performance

Why This Matters

Real-world mission scenarios involve the use of multiple systems of varying complexities and pedigrees working together to achieve the desired lethal effect. The emergence of highly network-centric concepts, greater dependency on connectivity, and the use of large amounts of data from a wide array of shooters and sensors across multiple domains, at machine speeds, warrants a review of our T&E processes within individual acquisition programs. Evaluating warfighting capability is further challenged by asynchronous updates and continuous evolution of the various components that comprise these system-of-systems operations. This demonstrates an inherent need to continually characterize the interoperability of such systems and their effectiveness as would be employed by the Combatant Commands. With the emergence of joint all-domain command-and-control solutions and the concept of kill webs, it is important to define the process and the required T&E tools that would effectively measure the success rates of mission threads, concepts, and solutions.

Drivers: Joint warfighting concepts, the attributes of the future Joint Force as defined in the National Defense Strategy 2022, advanced sensors, advanced communications and networks, cyber and electromagnetic spectrum warfare, AI/machine learning.

Enablers: Combatant Commands, requirements, requirements developers, large forces exercises, training exercises, digital technologies, AI/machine learning, data/knowledge management solutions.

Image Source: Lockheed Martin

✓ What Success Looks Like

DOT&E intends to establish a multi-disciplinary team comprised of government experts, consultants, academia, industry partners, small businesses, intelligence community and international partners to demonstrate progress in meeting the desired end state for Pillar 1. Specifically, in FY23, DOT&E will:

- Develop T&E concepts for evaluating kill webs/mission threads and joint all-domain command-and-control solutions.
- Provide a detailed assessment of existing large force exercises and a cost estimate to evaluate mission threads.
- Leverage TETRA and the Joint T&E Program to actively contribute to existing forums (e.g., Acquisition, Intelligence and Requirements) comprised of the intelligence community, Joint Staff, USD(A&S), USD(R&E) and the Services charged focused on scenario and mission thread approach. Visibility into such mission threads will inform T&E requirements needed to evaluate their operational effectiveness, suitability, survivability, and lethality.

Success will depend on collaboration with USD(R&E) and its Mission Engineering Team, the Combatant Commands, large force exercise leads, USD(A&S), the Intelligence Community, and existing Joint All Domain Command & Control cross-functional teams.

The table below highlights Key Actions and Deliverables that DOT&E has initiated to enable this Line of Effort. T&E enterprise initiatives will be subsequently identified.

	Key Action	Deliverable	FY23 Target Deliverable	FY24 Target Deliverable	FY25 Target Deliverable
1.2.1	Develop a T&E concept for evaluating joint all-domain command-and-control capabilities/kill webs/ mission threads while leveraging ongoing large force exercises.	DOD Manual defining T&E of mission threads, test resources, and anticipated limitations to adequately evaluate emerging joint all-domain command-and-control solutions/kill webs/mission threads.	Defined list of mission threads and best practices with demonstrated applicability to joint warfighting concepts	Drafted T&E strategy to test and evaluate select mission threads	Published DOD Manual defining T&E infrastructure, resources, and authorities to execute T&E mission thread strategy
1.2.2	Identify what new requirements and funding profiles would need to be established to enable the effective evaluation of mission threads.	Memorandum to Joint Staff and OSD Cost Assessment office summarizing new, mission-level requirements, the operating model for how such assessments would be executed, and a required funding profile.	Drafted key components of Memorandum	Published Memorandum	Coordination with the Joint Staff to implement the Memorandum.
1.2.3	Adequately represent the latest threats in testing as compared to the latest intelligence reports.	A data dashboard (linked to the range dashboard) that compares the availability of validated threat surrogates in test as compared to the operationally representative and relevant threats.	Scoped dashboard requirements	Developed Minimum Viable Product of dashboard	Finalized dashboard

Accelerate the
delivery of weapons
that work

Pillar 2

2.1 Develop and implement an enterprise-level T&E data management solution

Why This Matters

Data are a strategic asset that fuel automation and algorithms designed to alleviate our workload, speed up our processes, help us achieve new insights, and achieve T&E at scale and speed. As data-driven and complex systems continue to proliferate, it is important to develop T&E data and interface standards, stores, and platforms to ensure that the data are credible, trustworthy, available, and secure across the T&E enterprise. The T&E community must demonstrate their compliance with and contribution to the DOD Data Management Strategy.

Drivers: Adaptive acquisition framework, delivery of warfighting capability at speed of need, advanced technologies with complex performance metrics, increased attack surface.

Enablers: Digital technologies, automation, AI/machine learning, advanced data analytics, knowledge management solutions.

✓ What Success Looks Like

DOT&E intends to establish a multi-disciplinary team comprised of government experts, consultants, academia, industry partners, small businesses, intelligence community and international partners to demonstrate progress in meeting the desired end state for Pillar 2. Specifically, in FY23, DOT&E will:

- Develop OT&E and LFT&E data standards to support knowledge management and interoperability needs.
- Inform the development of an easy-to-use application through which users can search for T&E documents and gain desired, high-level analytical insights.

Success will depend on the cooperation of Service Test Agencies, ranges, and Program Offices to inform the usability and utility of proposed data/knowledge management (KM) solutions. Success will also depend on close coordination with the TRMC as the office of primary responsibility for this line of effort, as well as CDAO and USD(A&S) to ensure interoperability with ADVANA and other contracting efforts.

The table below highlights Key Actions and Deliverables that DOT&E has initiated to enable this Line of Effort. T&E enterprise initiatives will be subsequently identified.

	Key Action	Deliverable	FY23 Target Deliverable	FY24 Target Deliverable	FY25 Target Deliverable
2.1.1	Delivery of adequate data repositories, data management platforms, and analysis environments to T&E ranges, test agencies, and acquisition programs that allow for data ingestion and data analytics in near real-time across multiple security levels.	OT&E and LFT&E data standards and requirements for data stores, to include intellectual property rights.	Drafted data standards and requirements for DOT&E KM platform that can inform T&E enterprise solution	Completed enterprise review of OT&E and LFT&E standards and requirements	Published OT&E and LFT&E standards and requirements
		Pilot activities at select ranges, test agencies, and/or acquisition programs to modernize their data storage, management, and automated analysis prototype suites.	Commenced pilot activities	Completed pilots at select number of sites or programs	Scaled pilots
2.1.2	Implement smart word processing, document insights, and executive analytics capabilities for use in test planning, report development, and aggregate report analyses.	Tool[5] that enables document discovery and metadata management including smart document generation and executive analytics.	Developed tool – unclassified	Developed tool – classified	A process to proliferate the solution across the T&E enterprise, as needed.
2.1.3	Implement industry best practices for secure authentication, access management, encryption, monitoring, and protection of T&E data at rest, in transit, and in use.	DOD Manual summarizing best practices and recommendations for implementing and adopting these practices as part of the actions described in 2.1.1 and 2.1.2.	Developed initial material for DOD Manual; identified core principles.	Initiated investigation of Zero Trust Architecture	Completed investigation of Zero Trust Architecture, incorporated in Manual

---

[5] Tool is defined as *fully functional software*

Why This Matters

Modern model-based engineering and adaptive inference processes offer integrated, holistic approaches to generating and managing knowledge of system performance throughout the lifecycle. Early test data from system components, for example, can be integrated into a larger system model to predict mission-level performance early in development. Advanced performance inference techniques (e.g., Bayesian or similar) can be used to carry forward data from early prototypes through evaluation of production-representative systems. Moreover, model-based engineering can eliminate manual workflows through automation that enables generation and distribution of up-to-date dynamic reports on systems and their status in the acquisition life cycle.

Drivers: Adaptive acquisition framework, Shift Left or integrated T&E approaches, knowledge management, early problem discovery, large volume of disparate but related data.

Enablers: Digital technologies, automation, AI/machine learning, Bayesian or similar methods, behavioral learning methods, system modeling language (SysML), relational and non-relational databases.

✓ What Success Looks Like

DOT&E intends to establish a multi-disciplinary team comprised of government experts, consultants, academia, industry partners, small businesses, intelligence community and international partners to demonstrate progress in meeting the desired end state for Pillar 2. Specifically, in FY23, DOT&E will:

- Leverage unique expertise in Bayesian and similar methods to support the development of pilot programs that include both retrospective and forward-looking snapshots of what an integrated T&E program looks like at each phase in the program life cycle.
- Develop an example of a model-based T&E Master Plan using SysML, Structured Query Language databases, or other data model-backed technologies.

Success will depend on the cooperation of Program Executive Offices for the pilot programs, in particular their sharing of pertinent data and ensuring alignment on the proposed solutions. Success will also depend on close coordination with the Director, DTE&A and the digital engineering group within USD(R&E), as well as USD(A&S) and the Services to accelerate the use of model-based engineering in acquisition programs.

The table below highlights Key Actions and Deliverables that DOT&E has initiated to enable this Line of Effort. T&E enterprise initiatives will be subsequently identified.

	Key Action	Deliverable	FY23 Target Deliverable	FY24 Target Deliverable	FY25 Target Deliverable
2.2.1	Accelerate the development of tools that enable adequate performance inference from a growing body of evidence.	A retrospective exemplar in the form of a prototype Bayesian Decision Support System for performing integrated, sequential T&E.	Developed prototype of Bayesian Decision Support	Piloted Bayesian Decision Support solution	Scaled solution for additional programs
2.2.2	Accelerate the development of solutions that enable digital representations of T&E requirements and program tracking as compared to system design and operational performance requirements.	Digital model prototypes and training for two acquisition programs, to include digital Integrated Decision Support Keys.	Initiated pilot of DevSecOps framework and development of IDSK and model prototypes	Completed pilot and solution development, and initial scaling of solutions	Scaled solution for additional programs
2.2.3	Support the development of test-driven systems engineering methods that elucidate system deficiencies—and how to correct them—early in system development life cycles.	Computational framework that enables early deficiency discovery and correction for mission engineering.	Developed initial framework	Validated and finalized framework	Scaled solution for additional programs.

Improve the survivability of DOD in a contested environment

3.1 Standardize and automate mission-based risk assessments

Why This Matters

Seamless integration of multiple systems and technologies working together across multiple domains introduces a potential for vulnerabilities that cannot be evaluated one system or one threat at a time. As discussed in the first (Test the Way We Fight) pillar, testing must consider the mission thread, specifically the composition of weapon systems, networks, critical infrastructure, equipment, and Tactics, Techniques, and Procedures, in accordance with the Defense Planning Guidance, to fully account for the attack surface. A rapid and accurate mission-based risk assessment would define specific steps to enhance mission assurance and identify the defenses required against threats to those missions.

Drivers: Multi-domain operations, continuously changing and dynamic operating environment, exponentially increasing attack surface and vectors, synergistic threat effects.

Enablers: System Theoretic Process Analysis, digital technologies, AI/machine learning, modern software engineering.

✓ What Success Looks Like

DOT&E intends to establish a multi-disciplinary team comprised of government experts, consultants, academia, industry partners, small businesses, intelligence community and international partners to demonstrate progress in meeting the desired end state for Pillar 3. Specifically, in FY23, DOT&E will:

- Leverage Joint Technical Coordinating Group for Munitions Effectiveness (JTCG/ME) to develop a concept for full-spectrum survivability evaluation that includes agile verification and validation, and agile representation of threats in test and training.
- Initiate the development of credible digital technologies that will enable mission-based risk assessments and the full-spectrum survivability concept.

Success will depend on coordination with Program Offices to share relevant data and inform the proposed digital technology solutions.

The table below highlights Key Actions and Deliverables that DOT&E has initiated to enable this Line of Effort. T&E enterprise initiatives will be subsequently identified.

	Key Action	Deliverable	FY23 Target Deliverable	FY24 Target Deliverable	FY25 Target Deliverable
3.1.1	Accelerate the development of credible digital technologies to enable adequate and efficient characterization of system designs, prioritization of vulnerabilities, potential attack engagement conditions, and evaluation of threat effects on the mission.	A tool or set of integrated tools capable of predicting vulnerabilities and their mission effects when facing kinetic and non-kinetic threats	Scoped concept and requirements	Developed tool/set of tools Minimum Viable Product	Delivered tool/set of tools capable of accommodating continuous system and threat evolution
		A concept for infrastructure that will enable dynamic/agile threat updates.	Scoped infrastructure concept	Drafted infrastructure concept	Support the implementation of the infrastructure concept
		Documented process describing key measures, integrated test, and evaluation design to meet those measures, test resources and anticipated limitations to adequately evaluate mission-level vulnerabilities.	Published Cyber Survivability Whitepaper	Develop and implement process changes for other operationally relevant threats.	Identify process by which results are being proliferated to the warfighter and decision makers.

3.2 | **Emphasize cyber and electromagnetic spectrum survivability**

Why This Matters

The weapon systems of today and the future are defined by both software and hardware. Battle networks are central to the kill web, and information technology is at the heart of cyber, space, and electromagnetic spectrum warfare. The complex interactions between software and hardware can sometimes be difficult to predict or evaluate. Our challenge is to evaluate cyber-physical systems against advanced cyber and electromagnetic spectrum threats at scale and speed. Attack surfaces are growing exponentially, reaching into supply chains, software factories, and pipelines; the electromagnetic spectrum; and an array of cloud solutions. We therefore must aggressively pursue verified and validated digital tools and transformative technologies to manage cyber, electromagnetic spectrum and advanced kinetic threat survivability T&E and assess the effectiveness of countermeasures and other self-defense solutions.

Drivers: Cyber aggressors; contested, congested, and constrained electromagnetic spectrum environment; dynamic and constantly evolving threats; larger attack surface.

Enablers: Digital technologies, automation, AI/machine learning, data, knowledge management solutions.

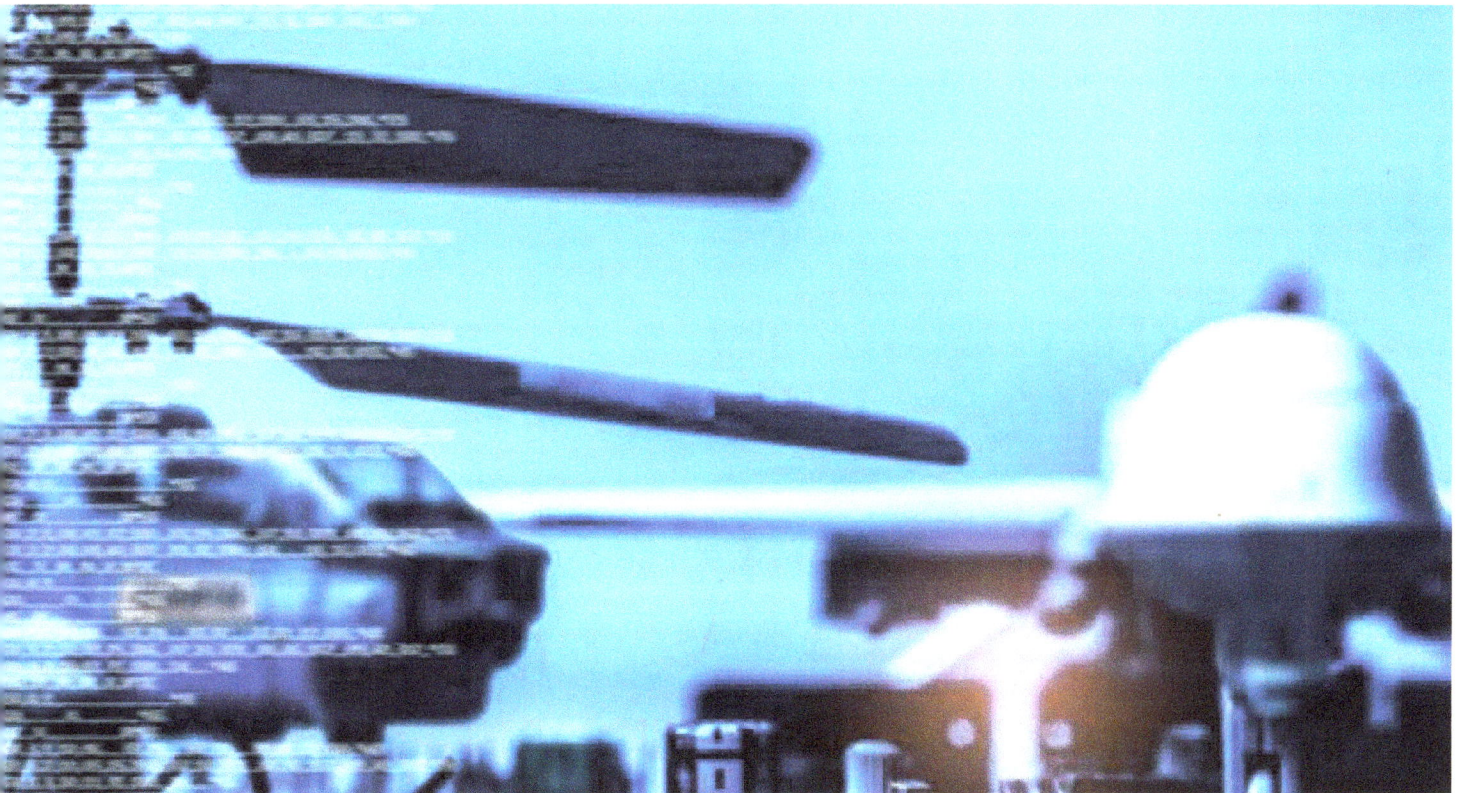

✓ What Success Looks Like

DOT&E intends to establish a multi-disciplinary team comprised of government experts, consultants, academia, industry partners, small businesses, intelligence community and international partners to demonstrate progress in meeting the desired end state for Pillar 3. Specifically, in FY23, DOT&E will:

- Leverage TETRA, JTCG/ME – Joint Live Fire (JLF) program, and Joint Aircraft Survivability Program Office (JASPO) to accelerate the delivery of accurate yet agile and dynamic representations of contested cyber- and electromagnetic-spectrum environments and advanced kinetic threats for use in testing
- Support development of automation processes in the cyber domain and to identify gaps in cyber and electromagnetic spectrum T&E.
- Accelerate progress in development of digital tools necessary to accurately evaluate the ability to detect and recover from cyber and electromagnetic spectrum threats.

Success will depend on the shared insight of Service Test Agencies and Program Executive Offices to align on identified gaps and converge on proposed solutions. Success will also depend on close coordination with Intelligence Community, the TRMC and the digital engineering experts within USD(R&E) and the Services.

The table below highlights Key Actions and Deliverables that DOT&E has initiated to enable this Line of Effort. T&E enterprise initiatives will be subsequently identified.

	Key Action	Deliverable	FY23 Target Deliverable	FY24 Target Deliverable	FY25 Target Deliverable
3.2.1	Develop a process to enable continuous identification, digital documentation, prioritization and tracking of opportunities and new gaps in cyber and electromagnetic spectrum T&E, in coordination with actions 1.1 and 1.2.	Requirements data dashboard accessible by programs, T&E community, and range stakeholders.	Identified data inputs and scoped requirements	Developed Minimum Viable Product of dashboard	Finalized dashboard
		Process for collecting and prioritizing T&E capability requirements.	Defined initial process	Finalized process	Refined process
		Deliver tool(s) to measure effectiveness of cyber defensive technique	Identified data inputs and scoped requirements	Developed Minimum Viable Product of tool(s)	Finalized tool(s)

3.3 Evaluate operational performance in a contested space environment

Why This Matters

Space is increasingly congested and highly contested, with a broad array of rapidly evolving threats. Reliance on space-based capabilities has sharpened DOD's – and our adversaries' – focus on deploying both offensive and defensive weapons in space. Because the DOD must operate in this contested environment, the T&E enterprise must be ready to accurately evaluate space-based and space-dependent systems' operational performance, to include survivability against current and anticipated threats.

Drivers: Precision navigation and timing; intelligence, surveillance, and reconnaissance; command and control provided by communications and observational satellites; proliferation of kinetic and non-kinetic threats to space assets.

Enablers: Digital technologies, automation, AI/machine learning, data, and knowledge management solutions.

✓ What Success Looks Like

DOT&E intends to establish a multi-disciplinary team comprised of government experts, consultants, academia, industry partners, small businesses, intelligence community and international partners to demonstrate progress in meeting the desired end state for Pillar 3. Specifically, in FY23, DOT&E will:

- Leverage TETRA to deliver accurate representations of the contested environment and an assessment of our ability to replicate that environment in test.
- Develop the T&E concept for testing systems in a contested space environment.

Success will depend on close coordination with TRMC and DTE&A within USD(R&E), as well as Service T&E Agencies to include the Space Force.

The table below highlights Key Actions and Deliverables that DOT&E has initiated to enable this Line of Effort. T&E enterprise initiatives will be subsequently identified.

	Key Action	Deliverable	FY23 Target Deliverable	FY24 Target Deliverable	FY25 Target Deliverable
3.3.1	Define the threats to space-based systems and a process for evaluating operational performance of DOD systems in a contested space environment and the performance of space systems.	Accurate description of the contested space and our ability to represent that environment in test and training.	Scoped requirements, including M&S capabilities	Defined process to validate requirements, including M&S capabilities	Defined process to integrate requirements into the evaluation of space-based systems
		DOD Manual describing key measures, integrated T&E design to meet those measures, test resources, and anticipated limitations to adequately evaluate the performance of DOD systems in a contested space environment.	Drafted initial key elements of manual	Developed guidance for testing of space systems and systems in a contested space environment	Published Manual and implemented in acquisition programs

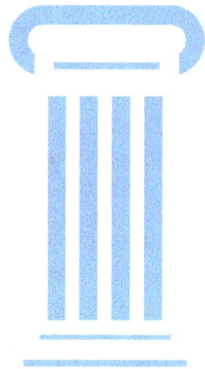

Pioneer T&E of weapon systems built to change over time

Pillar 4

4. Increase the use of credible digital twins in T&E

Why This Matters

The combination of new domains and operational constraints makes verified, validated and accredited digital technologies the necessary, practical approach for development and T&E of certain systems where live fire T&E is not possible or practical. For example, digital twins that can be subjected to repeated cyber-attacks – as the system itself, the threats it will face, and adversary tactics, techniques, and procedures change over time – will help developers and program managers improve system cyber survivability at the required pace. A digital twin is a high-fidelity digital representation of a physical object. These types of models allow us to find out how real-world objects might behave under different conditions or requirements. The defining feature of a digital twin is the ongoing data integration between the digital model and its physical unit counterpart. Digital twins have begun to incorporate transmission of real-time data sensed by the real-world object. These new, higher-resolution sensor data allow the digital twin to reason about future behaviors, then transmit feedback to the physical object. This could be particularly useful in enabling continuous monitoring of operational performance of systems as they evolve over time. While digital twins create new opportunities for T&E to determine the performance of continuously evolving systems, they also create new verification, validation, and accreditation challenges.

Drivers: Complexity of emerging technologies, threats, and the operating environment; the dynamic nature of the threat; the Adaptive Acquisition Framework; speed to field.

Enablers: Digital technologies, AI/machine learning, data/knowledge management solutions.

✓ What Success Looks Like

DOT&E intends to establish a multi-disciplinary team comprised of government experts, consultants, academia, industry partners, small businesses, intelligence community and international partners to demonstrate progress in meeting the desired end state for Pillar 4. Specifically, in FY23, DOT&E will:

- Lead a workshop to capture current state and needs, define a research agenda, and develop methods/criteria to effectively leverage the use of digital twins concepts in T&E.
- Leverage TETRA initiatives to demonstrate the ability of the intelligence community to use digital twins in representing the threat.

Success will depend on close coordination with the digital engineering group within USD(R&E) and the Services, USD(A&S), the intelligence community and Program Offices.

The table below highlights Key Actions and Deliverables that DOT&E has initiated to enable this Line of Effort. T&E enterprise initiatives will be subsequently identified.

	Key Action	Deliverable	FY23 Target Deliverable	FY24 Target Deliverable	FY25 Target Deliverable
4.1.1	Develop a process and criteria that need to be met to effectively leverage the use of digital twins in T&E to include an agile and iterative verification, validation, and accreditation process to continuously improve and update digital twins across the program lifecycle as new live data become available.	An addendum to the DOD Manual on modeling and simulation that describes the effective use of digital twins in T&E and the associated verification, validation, and accreditation process.	Identified DOD case studies and drafted M&S companion guide	Published Manual	Continuous evaluation of the credibility of digital twins in acquisition programs for the purposes of T&E.

Evaluate the operational and ethical performance of AI-based systems

Why This Matters

AI-based systems have accelerated the need to re-engineer T&E to enable continuous assessment even while fielded. The T&E enterprise must monitor and evaluate the drift in deployed AI models' behavior, which could occur when real-world data deviate from the training data used to create the model. Testing also must demonstrate, with confidence, that AI-based systems are responsible, ethical, equitable, traceable, reliable, and governable. Ethical and safe use of AI is necessary to reduce risks to U.S. strategic initiatives, reputation, operations, legal standing, and privacy issues. Due to their reliance on data, however, AI-based systems are uncertain by nature. Emerging approaches that have the potential to address such uncertainty propagation deserve further investigation.

Drivers: Modern software engineering, DOD Responsible Artificial Intelligence Strategy, joint all-domain command-and-control solutions, complex operating environments, cognitive threats.

Enablers: Digital technologies, data/knowledge management solutions.

✓ What Success Looks Like

DOT&E intends to establish a multi-disciplinary team comprised of government experts, consultants, academia, industry partners, small businesses, intelligence community and international partners to demonstrate progress in meeting the desired end state for Pillar 4. Specifically, in FY23, DOT&E will:

- Facilitate innovation brainstorming sessions to lead and support the three actions outlined below.

Success will depend on close coordination with TRMC and DTE&A within USD(R&E), the Chief Digital and Artificial Intelligence Office, and Service T&E Agencies and pockets of excellence in this domain.

The table below highlights Key Actions and Deliverables that DOT&E has initiated to enable this Line of Effort. T&E enterprise initiatives will be subsequently identified.

	Key Action	Deliverable	FY23 Target Deliverable	FY24 Target Deliverable	FY25 Target Deliverable
4.2.1	Identify areas that require further research to support adequate operational and ethical performance of AI-enabled systems (e.g., research/develop procedures for translating AI-enabled system testing into verifiable requirements that incorporate DOD AI ethical principles).	Ontology of research needs summarizing the milestones and deliverables needed to evaluate the operational and ethical performance of AI-enabled systems, to include the requirements and processes for AI red teams.	Defined use case of evaluation of AI-based system, identified gaps, and scoped requirements	Defined action plan to address gaps in evaluation and allocate resources accordingly	Implementation of the research agenda with pilot/demonstration case intended to close the identified T&E gaps.
4.2.2	Develop policies, standards, metrics, and a risk-based framework for assessing performance of emerging AI-enabled systems.	DOD Manual and accompanying guidance clarifying policies, standards, metrics, & a risk-based framework for assessing performance of emerging AI-enabled systems.	Drafted Manual	Codified guidance with demonstrated application to AI-enabled program	Updated Manual, as required, with applications to real acquisition programs.

4.2 Advance the evaluation of software-reliant systems' operational performance

Why This Matters

Modern warfighting systems are increasingly software-reliant. They are developed through complex software pipelines filled with myriad tools intended to ensure automatically that the product is effective and secure. Yet, developers more and more frequently utilize open source and third-party software, which raises risk from the security and sustainability perspectives.

It is important to identify new approaches to address change propagation within software-reliant systems. The T&E community needs to influence and measure the development and cyber defense of software pipelines and factories upfront with accredited tools, techniques, and procedures. We need to embrace automated testing at every level and ensure that our rigorous standard of testing continues to occur at the speed of relevance.

Drivers: Modern software engineering, complex technologies and threats, dynamic operating environment.

Enablers: Digital technologies, AI/machine learning, data/knowledge management solutions.

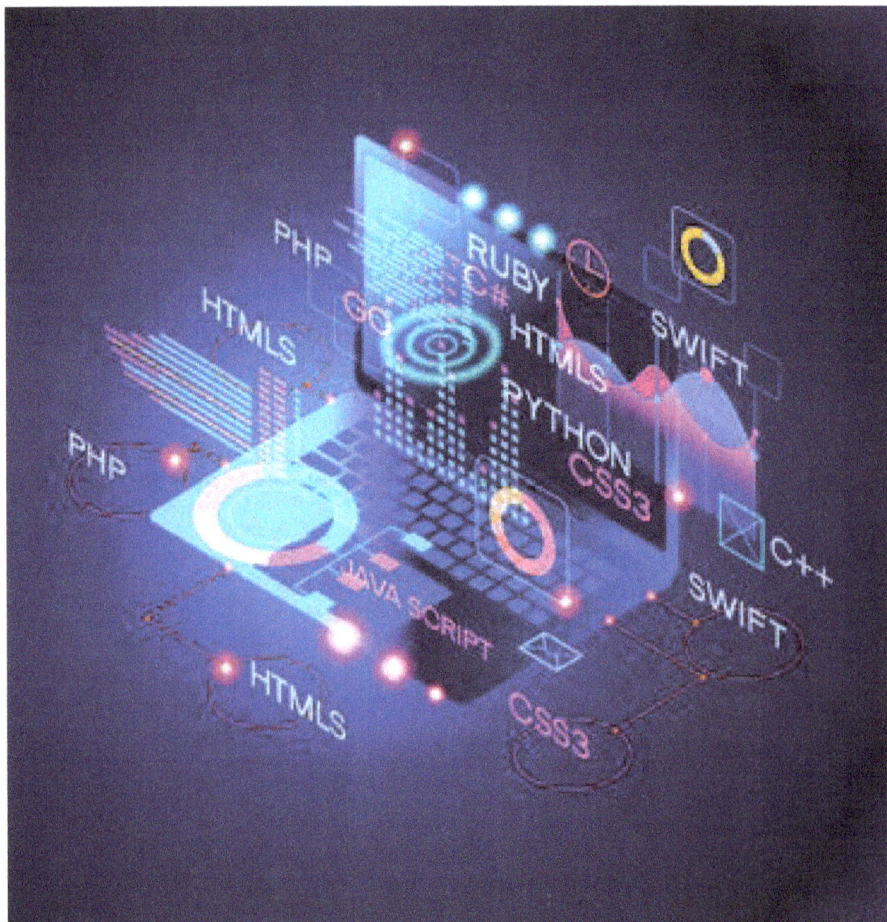

✓ What Success Looks Like

DOT&E intends to establish a multi-disciplinary team comprised of government experts, consultants, academia, industry partners, small businesses, intelligence community and international partners to demonstrate progress in meeting the desired end state for Pillar 4. Specifically, in FY23, DOT&E will:

- Define a rapid, iterative process for software OT&E that provides continuous insight into ongoing effectiveness, suitability, and survivability.

Success will depend on close coordination with the DTE&A within USD(R&E), Service Test Agencies, USD(A&S), CDAO, DOD CIO, and the Program Offices.

The table below highlights Key Actions and Deliverables that DOT&E has initiated to enable this Line of Effort. T&E enterprise initiatives will be subsequently identified.

	Key Action	Deliverable	FY23 Target Deliverable	FY24 Target Deliverable	FY25 Target Deliverable
4.3.1	Develop guidance for how to conduct continuous functional, operational, and cybersecurity T&E of software factories to support ongoing effectiveness, suitability, and survivability assessments.	Update to the Companion Guide for Software T&E describing the process for maintaining a proper balance among software pipeline features, defensibility, and stability.	Updated Companion Guide	Coursework and examples.	Application to acquisition programs.
		Minimum essential data elements and best practices for bill of material (BOM) generation. Pilot BOM pipeline, minimally supporting data capture/analysis/strategic decision support.	Identified data elements and best practices	Coursework and examples.	Application to acquisition programs.
		Guidance for using advanced tools in support of OT test generation.	Published guidance	Coursework and examples.	Application to acquisition programs.
4.3.2	Identify, investigate, and pilot automated T&E tools into support of shifting OT&E data collection to left and right to provide continuous insight into ongoing effectiveness, suitability, and survivability.	Software T&E tool directory that identifies an OSD-verified product stack with mapped capabilities, use cases, and lessons learned.	Scoped directory and established partnerships	Drafted and socialized directory	Finalized directory
		Established process for software T&E enabling end-to-end mission thread testing by utilizing traceability between mission thread testing & smaller epics'/user stories'	Surveyed problem	Established process	Application to acquisition programs.
4.3.3	Determine operational test considerations for emerging software-centric technologies and techniques.	Metric characterization, procedures, and assessment of offensive cyber capabilities.	N/A	Defined metrics, procedures, & assessment	Coursework and examples.
		Addendum to Companion Guide clarifying guidance for operational testing of 5G-enabled/affected systems.	N/A	Published addendum	Coursework and examples.
		Addendum to Companion Guide clarifying guidance for OT of Cloud systems.	N/A	Published addendum	N/A

Foster an agile and enduring T&E Enterprise Workforce

Pillar 5

5. **Identify and track T&E workforce competencies and capabilities**

Why This Matters

A structured approach for the development and sustainment of the T&E enterprise workforce will enhance workforce agility and response to emerging T&E requirements. The nature of T&E evolved drastically over the past decade, and the reemergence of Great Power conflict, necessitates a thorough review and refinement of the T&E Competency Model to ensure T&E professional learning and development aligns to the enterprise's needs. Coupled with modernized competencies, the enterprise should better track and manage the T&E workforce's overall readiness in real-time to deliver improved talent management initiatives which strengthen the overall enterprise and provide an optimal T&E professional talent experience. The T&E Functional Integration Team (FIT) is an established cross-enterprise body well positioned to lead these initiatives, and extra support will launch the competency model refinements and T&E workforce database in addition to the T&E FIT's existing initiatives.

Drivers: Emerging technologies and threats, DOD Data Management Strategy, Adaptive Acquisition Framework, data-driven workforce planning.

Enablers: Digital technologies, AI/machine learning, data/knowledge management solutions.

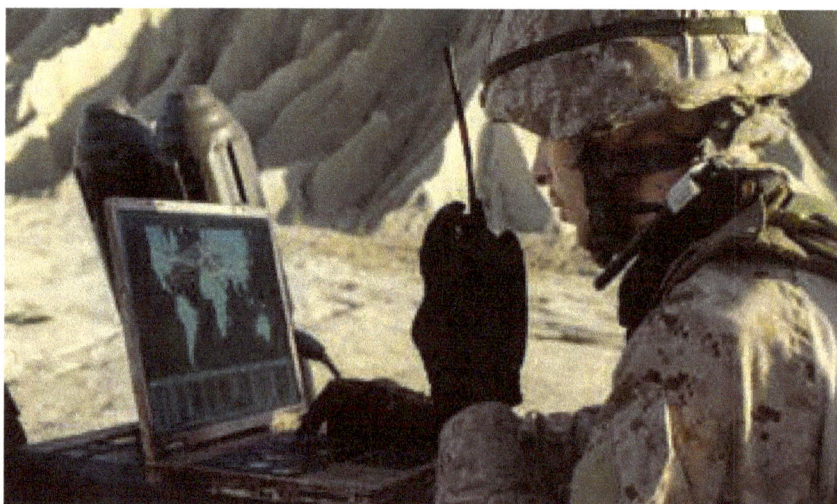

✓ What Success Looks Like

DOT&E works within the T&E FIT to establish a multi-disciplinary team comprised of government experts, consultants, academia, industry partners, small businesses, intelligence community and international partners to demonstrate progress in meeting the desired end state for Pillar 5. Specifically, in FY23, DOT&E will:

- Engage Operational Test Agencies (OTAs) and Live Fire Test and Evaluation Organizations to identify emerging knowledge, skills, and abilities (KSAs), crosswalk KSAs with the T&E Competency Model to identify potential areas of improvement, and submit competency model refinement recommendations to the T&E FIT

Success will depend on close coordination with personnel and readiness teams within USD(A&S) to include the Defense Acquisition University, USD(R&E), USD(P&R), Service Test Agencies, and Program Offices.

The table below highlights Key Actions and Deliverables that DOT&E has initiated to enable this Line of Effort. T&E enterprise initiatives will be subsequently identified.

	Key Action	Deliverable	FY23 Target Deliverable	FY24 Target Deliverable	FY25 Target Deliverable
5.1.1	Coordinate across T&E enterprise to define clear competency criteria for a T&E professional identifying and validating common capabilities needed to successfully conduct T&E.	Updated T&E professional competency model outlining core T&E practitioner knowledge, skills, and abilities.	Identified enterprise competency model recommendations	Updated T&E competency model	Established Process for maintaining updated T&E competency model
5.1.2	Identify T&E-specific skill codes to track available capabilities representative of all T&E practitioners	Updated database capturing T&E workforce functions, associated skills codes, and capabilities critical to the T&E mission.	Identified opportunities for DOT&E workforce to integrate into Acquisition Position Category Descriptions (PCD)	Identified PCD recommendations	Updated PCD

5.2 Assess and address critical T&E workforce professional development needs

Why This Matters

T&E professionals of the future require access, bandwidth, and clear requirements to engage in continuous learning opportunities. Providing these will better prepare them for advances in T&E operational and technical capabilities needed to perform their duties. The T&E enterprise's current learning IT infrastructure is federated across numerous systems, which is not a bad feature; however, restricts the enterprise's ability to track and share relevant learning opportunities, track T&E professionals' learning records, and ultimately expand the types of learning opportunities easily accessible to T&E professionals. The T&E FIT is well positioned to lead and oversee the enterprise's push to better connect learning opportunities to the T&E workforce and track learning records and outcomes against the T&E Competency Model. The T&E learning apparatus should evolve as quickly as the T&E operating environment, resulting in easily adaptable courses, content, and training and workforce demands. Continuous learning can be better incentivized and accessible by leveraging the enterprise's currently federated learning data as a strategic asset, and overall gaps in learning may be more easily identified and addressed. While these gaps may identify courses to add or refine, they can also include more informal types of learning – ranging from mentorships and rotations to cross-T&E seminars and networking events.

Drivers: Emerging technologies and threats, DOD Data Management Strategy, Adaptive Acquisition Framework, data-driven workforce planning, industry competition.

Enablers: Digital technologies, AI/machine learning, data/knowledge management solutions.

Body

What Success Looks Like

DOT&E works within the T&E FIT to establish a multi-disciplinary team comprised of government experts, consultants, academia, industry partners, small businesses, intelligence community and international partners to demonstrate progress in meeting the desired end state for Pillar 5. Specifically, in FY23, DOT&E will:

- Establish core T&E learning needs in partnership with T&E FIT representatives, identify opportunities to expand the T&E enterprise's learning and professional development capabilities to support continuous learning for T&E professionals and improve professional development and retention.

- Expand scholarships/internships and job rotations intended to accelerate the delivery of new skills to the T&E field.

Success will depend on close coordination with personnel and readiness teams within USD(A&S) to include the Defense Acquisition University, USD(R&E), USD(P&R), Service Test Agencies, and Program Offices, in addition to relevant academic and industry partners.

The table below highlights Key Actions and Deliverables that DOT&E has initiated to enable this Line of Effort. T&E enterprise initiatives will be subsequently identified.

	Key Action	Deliverable	FY23 Target Deliverable	FY24 Target Deliverable	FY25 Target Deliverable
5.2.1	Recommend solutions to address T&E workforce training and education gaps and needs.	Learning needs assessment to identify and evaluate T&E learning journeys and course curriculum to meet future workforce demands T&E	Identified DOT&E learning needs assessment recommendations	Developed scalable DOT&E Learning Strategy to build continuous learning program	Implementation of the learning strategy and iteration based on user feedback.
5.2.2	Complete a review of and catalog existing T&E training and education opportunities, recommending approaches to improve current learning infrastructure.	T&E course catalog with training and education courses aligned to core T&E competencies and tracked via a single, easily accessible, and searchable course catalog.	Identified DOT&E workforce-focused training and education opportunities	Developed DOT&E workforce-focused digital course catalog	Completed T&E Enterprise Course Catalog
		Recommendations to advance the delivery of T&E professional development programs to promote a more engaging experience.	Identified instructional design recommendations for DOT&E AO Course and T&E Companion Guide trainings	Developed new and refreshed DOT&E course offerings	Updated DOT&E course offerings
5.2.3	Leverage existing and establish new experiential learning opportunities to access and build required expertise across the T&E enterprise.	Scholarship and temporary assignment opportunities within the T&E community using the Software and Cyber Innovation Center.	Drafted T&E Software & Cyber Innovation Center Charter and initiated implementation of standup activities	Implemented Virtual Innovation Center	Completed Innovation Center launch and scale activities

I-Plan Operating Model

The Operating Model below depicts the significance of collaboration across the array of stakeholders who drive and affect the future of the T&E enterprise. DOT&E is not positioned, nor would it want, to implement this I-Plan in isolation from the T&E enterprise and other relevant stakeholders. DOT&E is also not directing any of the actions in this document. This I-Plan identifies what are perceived to be common shortfalls that, if addressed soon, will help the T&E enterprise meet the demands of the future. The success of these actions depends on the participation of T&E community, their willingness to freely share information, and contribute as able to expedite the delivery of the required T&E capabilities. The model below identifies layers of governance, as well as key stakeholders, to demonstrate how the T&E community can align on and execute against the five strategic Pillars.

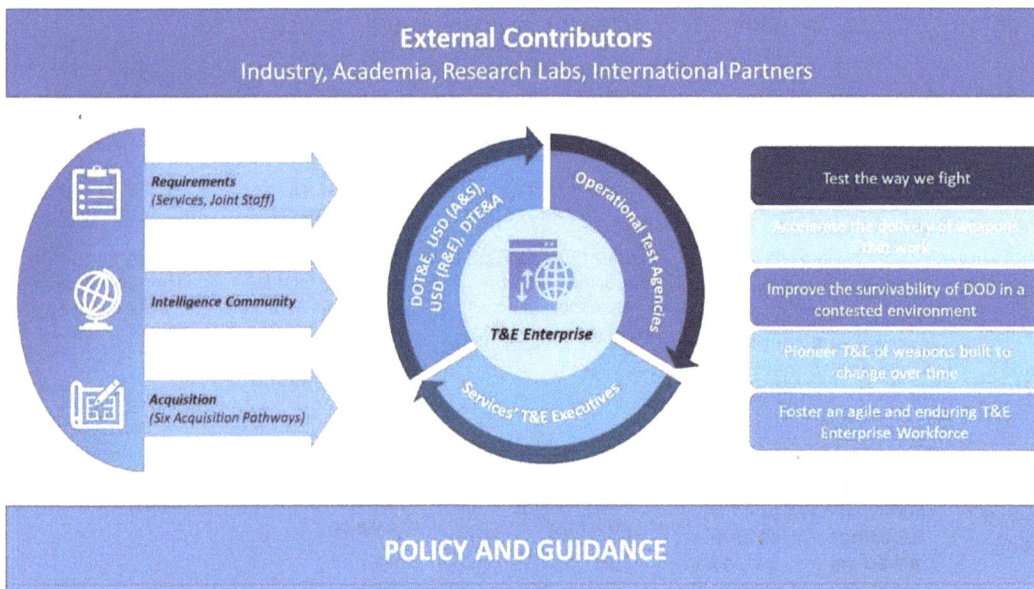

Requirements, intelligence, and the acquisition pathways drive the T&E process, as depicted above. Changes in capabilities, such as kill webs, complex all-domain environments, and gaps newly identified by intelligence reports, will steer acquisition decisions and a commensurate T&E response. Based on the requirements, intelligence, and mandates sourced from the six acquisition pathways, the T&E community will collaborate to identify and develop the T&E capabilities necessary to test and evaluate systems in the acquisition pipeline. These T&E activities will realize the goals of the five strategic pillars that will in turn inform T&E policy and guidance with potential to inform operational and system requirements, system development, and acquisition contracts.

Several external contributors are considered a critical part of the operating model. Industry, Academia, Research Laboratories, and International Partners may have conducted research and developed capabilities that the T&E community could leverage to accelerate progress.

Conclusion

The Department faces a shifting threat landscape and the need to swiftly leverage advanced technologies to increase the lethality, suitability, resiliency, survivability, agility, and responsiveness of our future Joint Force. To continue to deliver credible warfighting capability at the speed of need, we in the T&E enterprise must rethink the way we do business. We must become more agile, efficient and effective to adequately account for the technology disruptors as we face an inflection point in the scope, scalability, and capabilities of our infrastructure, tools, processes and workforce.

This I-Plan drives the T&E enterprise into the future through a series of coordinated enterprise-wide activities designed to achieve the future vision that is embedded in the strategy. It sets the framework to leverage ongoing government-based activities, the best practices of industry, academia and our allies to develop a future-ready T&E Enterprise. Our immediate next step is to establish a governance charter and launch multi-disciplinary teams for each of the five pillars with a clear direction on what they need to accomplish, when, why, and with what resources.

The T&E enterprise of the future will be agile, motivated by scenario/mission thread approaches, joint warfighting concepts and the power of digital tools and technologies. It will be strengthened by the effect of these changes on our ability to support the warfighter. It will be empowered by continuous learning and supported by unbound access to state-of-the art skills and technologies.

By implementing this strategy, the T&E enterprise will be better positioned to stay ahead of the adversary and continue to advocate for the warfighter and its mission as defined by the National Defense Strategy 2022.